Mr. Page's class was going to the Bridgestone Zoo. On the way, the school bus had to cross the Bridgestone Bridge.

The mighty bridge seemed to
shimmer in the bright morning light.

"I've heard of this bridge," said
Ginger. "My uncle is an artist. He
made a greenish painting of it."

The bus had to stop and wait at
the edge of the Bridgestone Bridge.

"This bridge has two large parts," said Mr. Page. "When the parts are up high, a large ship can pass under the bridge."

At the zoo, Mr. Page passed out little notebooks. He said, "Jot words on these pages to tell what you see. When we get back, you can write a sentence or two."

Ginger gave her friend a nudge and said, "I'll fill these pages right up!"

The class enjoyed many sights.
One big seal would dodge two little
seals as they swam. A golden eagle
sat high up on a rocky ledge. Its
home was inside a huge cage.

On the way home, the school bus
crossed the Bridgestone Bridge again.

"Look!" said Ginger. "The bridge's
lights sparkle like gems! Now I know
what I'll write."